To: Steve
Christmas 1997
Love,
Mom + John

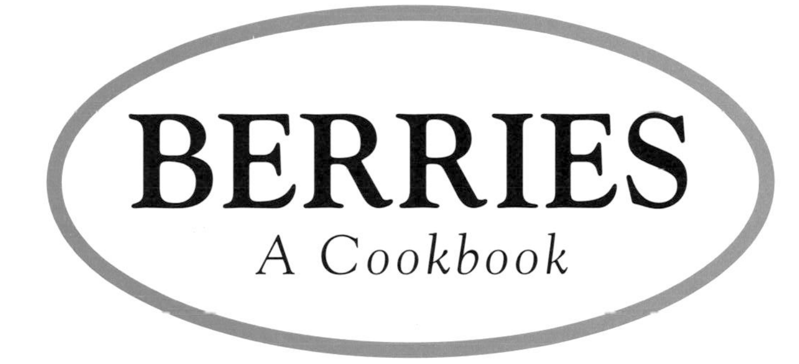
BERRIES
A Cookbook

BERRIES

A Cookbook

CHARTWELL
BOOKS, INC.

A QUANTUM BOOK

Published by Chartwell Books
A Division of Book Sales Inc
114 Northfield Avenue
Edison, New Jersey 08837
USA

This edition printed 1997

ISBN 0-7858-0787-X

QUMBBK

This book was produced by
Quantum Books Ltd
6 Blundell Street
London N7 9BH

Printed in Singapore by Star Standard Industries Pte Ltd

Acknowledgments

To my mother, my grandparents and Michelle.

Special thanks to Mary Trasko, Mildred Raucher and Mary Forsell for their contributions of materials for this book. Thanks to Michelle Hauser, Nancy and Sam Freitag, Loretta and Bruno Hauser, Kate Struby, Meggin Siefert, Robin Page, Maggie Jones, Judy Devine, Myriam Zwierzinska, Doug Hay, Rick Silverness, Jennifer Barnaby, prop stylist, and Brendan Mullany, photo assistant. Thanks to Marta Hallett, Ellen Milionis, Sarah Kirshner, Lindsey Crittenden and all the other support at Running Heads.

Thanks to the following for their generosity: ABC Bed Bath and Linens; Marek Cecula, designer, and Pascal Golay, designer; Contemporary Porcelain; Gear Stores; Bill Goldsmith, plate designer; Marble Connection; Umbrello.

THREE
Main Courses

FOUR
DESSERTS

INTRODUCTION

When I think of berries, the first thing that comes to mind is a sensation I experienced as a teenager. I was driving through central North Carolina with a friend on a sunny late morning in June. We were driving on a state highway when we found ourselves along a series of mountains, and decided to stop the car by the base of one of the mountains to hike.

Since there were no paths through the thickets, we had to create one. After a long climb we rested, and while catching our breath from the heat and the strenuous hike, we noticed a blackberry bush. We each dared the other to try a berry; one of us finally did, and when he didn't keel over from poison, and found that it was good, we managed to devour almost every berry on that bush.

Every time I eat a really fresh berry, I recall the combination of the sun, the fresh air, the mountains of North Carolina, being sixteen and the supreme sensual experience of the slightly sour berry juice bursting between my teeth. Even the word "berries" is enough to evoke the abundance of all these sensations.

It took me a long time and certain prejudices to overcome before I began cooking with berries; I didn't want to spoil their organic intensity. After all, cooking is by definition a process of destruction: breaking down fibers, altering flavors, changing forms. How could a fresh berry be improved? Cooking with berries demands different expectations and requires a different approach. Within each berry lies a multitude of different flavors and textures which can be drawn out to do whatever you as a cook want them to do. Depending on the demands of a dish, the elements within a berry can be used to thicken sauce, change color, balance spiciness, counteract bitterness – all while lending to the dish the distinct and sublime character of the particular berry. The idea of drawing out more savory and fuller, rounder flavors from a berry can be a very exciting challenge to an inquisitive cook.

The recipes in this book are designed to illustrate the diversity of cooking with berries. In some cases simply adding a fresh berry to an assortment of ingredients is all that is required, the pristine state of the berry speaking for itself within the arrangement of the dish. In other cases we put the berry through various cooking stages to elicit the different degrees of character in each berry. In fact almost entire identities change in different cooking processes.

For example, in comparing Blueberry Vichyssoise with Blueberry Pie, it's hard to believe the same ingredient is used in both dishes. The different tastes of the same berry in reaction to the other ingredients – one earthy, potatoey, the other syrupy, fruity and sweet – demonstrates two parts of the spectrum of this wide-ranging fruit.

Another striking example is what can happen to a soup when we substitute berries for other ingredients, Gazpacho calls for tartness, Usually a little vinegar or lemon juice fills that vacancy and gently supports the other ingredients – tomato, cucumbers, peppers – so that the focus of attention is on them, while the lemon or vinegar acts as a sort of post, quietly holding up the structure. In Cranberry Gazpacho, the cranberry is gentle enough not to overshadow everything else but intense enough to be the dominant flavor of the soup, with the other ingredients acting in support of it – an exciting variation on a traditional dish.

In a similar vein, I have chosen closely related recipes that share certain procedures and ingredients but use different berries. In Smoked Trout and Raspberry Salad with Lingonberry Dressing, raspberries are used for their soft flavor. A more sour berry would certainly work well with the bitter flavor of the arugula, but would confound the saltiness of the smoked fish. In the Bitter Greens Salad with Strawberries and Gooseberry Vinaigrette, strawberries are used so as not to detract from the sharpness of the goat cheese. The common element in both salads is the bitter flavor of the greens. Both berries work comfortably with their other accompaniments – the smoked trout and the cheese – and these accompaniments work equally well with the greens. What becomes interesting is the flavors that become apparent in the greens when combined with the berries. The bitterness seems to change when introduced to the exceptionally subtle differences in the two berries. The recipes illustrate the range of character of each berry and its capacity to work well with other ingredients in a way that a straight taste test between a strawberry and a raspberry could not.

The fragility of flavor that makes berries so desirable isn't always easy to obtain. You can't always be on a Carolina mountainside when blackberries reach their height of perfection. As with many other types of produce, different berries are in season at different times and do not grow for much of the rest of the year, nor do they grow in the most accessible places. Blueberries are in season in the Northern Atlantic States from August through September, and are in season in New Zealand through March to early April. Cranberries are in season in North America from early November through mid-February. Strawberries are in season in mid-June to July, raspberries are available May through June, but many berries are now available year-round. It is rare to experience a really fresh cloudberry at any time outside Scandinavia. Although modern shipping techniques can provide people all over the world with berries from other places, between the time delays of harvest and sorting, shipping and distributing, several days may have passed by the time a person in the United States goes to market for these berries. And although well-traveled berries remain fresh, they cannot provide a quintessential tasting experience.

When shopping for fresh berries, the main things to avoid are mold (fuzzy whiteness on the undersides of the berries) and softness. A little bit of mold is all right, but it spreads quickly. A berry is probably soft if it looks wrinkled or deflated. Moldy berries aren't good for cooking or eating. Soft berries, however, can be used for preserves or pies with no loss of flavor.

Generations of cooks have developed storage techniques that enable people to enjoy late-summer and spring berries in the middle of winter. Freezing is the most obvious method. Sealed airtight in a rigid plastic container or wrapped snugly in plastic wrap, berries will keep up to four or five months in the freezer.

Sun-drying is a natural means of preserving the essence of a berry. In the same way that a raisin is a dehydrated form of a grape, sun-dried berries retain much of their original berry essence. Gourmet and specialty food stores carry varieties of sun-dried berries. They can be stored at room temperature, in a cupboard or in a paper bag for months and months with no loss of sweetness or flavor. For several recipes in this book – Sun-dried Berry Bagels, for example – dried berries are used to a greater advantage than their fresh equivalent. The bagels undergo a wet cooking stage followed by a dry cooking stage. Dried berries maintain their form and withstand the drastic change in cooking techniques where the fresh varieties would lose their own moisture during the wet stage and bleed during the dry stage.

Wines, juices, brandies and vinegars also capture the essence of a berry in an extremely subtle way when used reservedly. The presence of a blueberry vinegar mixed with olive oil in a salad dressing can suggest aromatically the presence of a blueberry. The same blueberry vinegar can be reduced in a pan and mounted with butter to make a creamy sauce for fish. Depending on the vehicle – olive oil or butter – and the final subject of the dish – salad or fish – you can experience the essence of blueberry in two vastly different ways.

Perhaps my favorite means of saving berries is through preserves. Preserving melts the membranes that hold a berry together, breaks down the fine fibers that keep it from simply turning into juice and, through a long and slow cooking process, transforms a bunch of whole berries into something almost completely different. It is through this process that many interactions occur. Although a recipe for preserves is a simple series of steps, it is perhaps one of the most complex chemical exchanges of berry cooking. The membranes themselves seem to dissolve in the pot and evaporate with all the steam, but they remain, helping to thicken and preserve the berries.

The uses of preserves are as diverse and plentiful as the fresh berries themselves. When added to a reduced stock, they give it body and turn it into a sauce. When added to softened butter with egg yolk and fine sugar, they make the ultimate cake frosting. When added to sour cream they make a delicately balanced salad dressing. Or they can be enjoyed simply with sweet butter spread on good bread. For me, tasting well-made preserves is like being in the center of a berry.

For the most part, each recipe in this book – as is the case with most recipes – should be viewed as a blueprint to a dish, a guideline to use as a point of reference. Many of the berries are interchangeable. The "lumpy" berries – raspberries, blackberries, loganberries, boysenberries, mulberries – are natural substitutes for one another in shape as well as taste, as are the "smooth" berries – blueberries, cranberries, currants, lingonberries, gooseberries and rosehips. These are not hard and fast categories for substitutions, just suggestions. Regional, less common berries fall into these categories as well. Try using raspberries or strawberries for cranberries and you'll become familiar with the enormous range of possibilities within the berry family. This explorative approach can be applied to other groups of ingredients as well, adding an entire new dimension to your cooking repertoire.

ONE
BREAKFAST AND BRUNCH

DRIED BERRY BAGELS

1 package dry active yeast
1 tablespoon sugar
1 cup warm water (about 110º F.)
4 cups high-gluten flour
1 teaspoon salt
2 eggs
2 tablespoons vegetable oil
1/4 cup sun-dried blueberries
1/4 cup sun-dried cranberries
2 tablespoons water
1 egg yolk

Makes 14 bagels
Preparation time: 2 1/2 hours

Preheat oven to 375º F. In a bowl, place yeast, sugar and warm water. Let stand for 5 minutes. Yeast will foam. If the water is too hot or too cold the yeast will not foam and the process must be repeated.

Add 1 cup of flour, salt and the 2 eggs to the mixture. Whisk together to incorporate flour. Gradually add the remaining flour. Incorporate it with your fingers. Continue to add flour until the dough loses its stickiness. Knead the dough for 5 minutes. This will activate the protein in the flour to make it rise, and give it elasticity.

Coat a large bowl with half the vegetable oil. Place dough in the bowl and cover with a damp cloth. Let it rise at room temperature for about 30 minutes.

Cut dough into 14 equal pieces and allow to rise another 20 minutes. Work the sun-dried berries into the dough pieces.

With your finger, poke a hole in the center of each piece, and form into the shape of a bagel. Heat a large pot of unsalted water to boiling. Lower to a simmer. Place the bagels in the simmering water for 3-4 minutes. Turn over and simmer for another 3-4 minutes. Coat a large baking pan with the remaining vegetable oil. Beat 2 tablespoons of water with the egg yolk. Brush each bagel with the egg wash and place on the baking pan. Bake for about 35 minutes or until golden brown.

BLACKCURRANT CRÊPES

2 eggs

1/4cup sugar

2 tablespoons sifted, all-purpose flour

1/4 cup milk

1/4 cup fresh black currants

2 tablespoons unsalted butter

1/4 cup ricotta cheese

Serves 8-10

Preparation time: 1 1/4 hours

In a bowl, combine eggs, 1/4 cup sugar and flour. Add the milk. For thinner crêpes, add more milk. Let sit 1 hour. In a 2-quart saucepan, simmer the black currants, just covered with water and 1/4 cup of sugar. Cook for about 2 minutes. Remove from the water and let cool.

In a 6-inch non-stick omelet pan, melt 1/4 teaspoon butter. When the butter begins to crackle, add 2 tablespoons of the crêpe batter, enough to evenly cover the pan. When it forms a solid coating on the bottom, about 1 minute, flip it. Continue cooking for a few seconds, and slide onto a large clean surface. Let cool. Repeat until all of batter is used. Spoon about 2 tablespoons of the ricotta cheese along the diameter of each crêpe. Spoon about 2 tablespoons of the black currants directly alongside the cheese. Roll each crêpe to form a log.

RASPBERRY PRESERVE

1 quart fresh raspberries

1 cup sugar

1 quart cold water

skins from 2 Granny Smith apples, tied securely in cheesecloth

In a 4-quart heavy saucepan (copper preferred), heat all ingredients to a boil, stirring often.

Reduce heat to the lowest possible heat and cook for 6 hours, or until mixture is firm, stirring frequently. Add more water gradually, if necessary. Cool, uncovered, in refrigerator. When completely cool, cover with plastic or in a seal-tight glass jar. Can keep up to several months.

Makes 500-750 ml
(1-1¼ pints) preserve

Preparation time: 10 hours

PECAN MUFFINS WITH MULBERRIES

1/4 cup pecans, finely chopped

1/4 cup all-purpose flour

2 tablespoons baking powder

1/4 cup sugar

1 egg

2 tablespoons unsalted butter, melted and cooled

1/4 cup warm milk

1 teaspoon vanilla extract

pinch of salt

1/4 cup dried mulberries, or any dried berries

Serves 4

Preparation time: 30 minutes

Preheat oven to 350º F. Combine pecans, flour, baking powder and sugar. Add egg, butter, milk, vanilla, salt and mulberries. Let stand 10 minutes.

Fill four cups of a non-stick muffin tray with the batter. Bake 8-10 minutes, or until a toothpick inserted in the center comes out clean.

CORNMEAL MUFFINS WITH MULBERRIES

1/4 cup yellow cornmeal

1/4 cup all-purpose flour

2 tablespoons baking powder

1/4 cup sugar

1 egg

2 tablespoons unsalted butter, melted and cooled

1/4 cup dried mulberries, or any dried berries

1/4 cup black poppyseeds

1/4 cup warm milk

pinch of salt

1 teaspoon vanilla extract

zest of 1 lemon

Serves 4

Preparation time: 30 minutes

Preheat oven to 350º F. Combine cornmeal, flour, baking powder and sugar. Add egg, butter and mulberries. In a separate bowl, soak poppyseeds in milk for 5 minutes. Add both ingredients to the batter.

Add salt, vanilla and lemon zest. Let stand for 10 minutes. Fill 4 cups of a non-stick muffin tray with the batter. Bake for about 8-10 minutes, or until a toothpick comes out clean when you poke it through the center.

BLUEBERRY CORNMEAL GRIDDLE CAKES

1 cup cornmeal
1 teaspoon salt
2 tablespoons honey
1 cup boiling water
1 egg
1/2 cup milk
2 tablespoons unsalted butter, melted
1/2 cup all-purpose flour
2 tablespoons baking powder
1/2 cup blueberries
2 tablespoons unsalted butter

In a bowl, combine cornmeal, salt, honey and boiling water. Mix until smooth. Let cool slightly.

Add the egg, milk and melted butter to the cornmeal mixture.

Sift in the flour and baking powder. Add the blueberries. Set aside for 10 minutes. Melt 1 tablespoon butter in a large heavy skillet over medium-high heat.

Pour 4 tablespoons batter into skillet for each pancake. When bubbles pop around the edges, and begin to form in the center of each pancake, flip and cook for another minute.

Serves 4
Preparation time: 30 minutes

BOYSENBERRY SYRUP

1 cup fresh boysenberries
1 1/2 cups water
1 whole cinnamon stick
1/4 cup molasses
1/4 cup light brown sugar
1 teaspoon vanilla extract

Place boysenberries, water and cinnamon stick in a small heavy saucepan, and cook over a low heat for about 20 minutes. Remove cinnamon stick and strain.

Add molasses, sugar and vanilla. Let cool.

Makes 250 ml (8 fl oz)
Preparation time: 40 minutes

GOLDEN RASPBERRY CRUMBCAKE

1 tablespoon unsalted butter

3/4 cup, plus 2 tablespoons all-purpose flour

1 tablespoon baking powder

1/2 teaspoon salt

1/4 cup sugar

1/4 cup unsalted butter, melted and cooled

2 tablespoons vanilla extract

2 eggs

2 cups golden raspberries

1/2 cup sour cream

CRUMB TOPPING:

3/4 cup light brown sugar

1/2 cup unsalted butter, cut into small pieces

1 1/2 cups all-purpose flour

Serves 6

Preparation time: 45 minutes

Preheat oven to 350º F. With tablespoon of butter, evenly coat a 7- x 9-inch baking pan.

Dust with 2 tablespoons flour. Shake out excess. Refrigerate. In large bowl, combine remaining 1/4 cup flour, baking powder, salt and sugar. Add the melted butter, vanilla and eggs. Mix until well combined. Fold in the raspberries and sour cream.

Pour batter into baking pan. Let stand 10 minutes. In a separate bowl, combine brown sugar and butter pieces. Add flour and mix until evenly incorporated.

Sprinkle evenly over cake batter. Bake for 30 minutes or until a toothpick inserted in the center comes out clean.

TWO
Appetizers, Soups and Salads

BLUEBERRY VICHYSSOISE

5 new potatoes, washed in cold water
4-5 scallions, white part only
3 cups chicken stock (See below)
1/2 cup sour cream
1 cup blueberries
1/2 cup milk

Serves 4
Preparation time: 1 1/2 hours, plus overnight chilling

Cut washed potatoes into quarters.

Cut scallions into 1/2-inch pieces. In a 1-gallon pot, heat chicken stock, potatoes and scallions to a boil. Lower to a simmer. Cook for about 30 minutes, until potatoes are soft.

Strain potatoes, scallions and stock through a sieve, and return to pot. Bring back to a simmer over a low heat.

Whisk in the sour cream. Add 1/4 of the berries and cook on low heat until the first berries split. Set aside remaining berries.

Remove from heat and chill overnight.

With a fork, mash the blueberry mixture and thin with the milk until desired consistency.

Garnish with the remainder of the uncooked berries.

CHICKEN STOCK

2 tablespoons vegetable oil
2 large carrots, thinly sliced
8-10 shallots, finely chopped
4 scallions, finely chopped
1 Granny Smith apple, quartered with skins and seeds
1/4 cup white domestic mushrooms, coarsely chopped
1 pound chicken wings
20-40 ice cubes, roughly 6 cups
12-15 whole black peppercorns
2-3 cloves of garlic, crushed
1 bunch parsley

Makes 2-3 litres (3 1/2-5 pints)
Preparation time: 4 hours

Heat oil in a 2-gallon pot. When the oil is hot add the carrots, shallots, scallions, apple and mushrooms. Cook for 10 minutes.

Add chicken wings, ice cubes, peppercorns and garlic. Cover all ingredients with cold water and top with parsley. Bring to a boil, and immediately lower to a simmer. Cook for about 2 1/2-3 hours, skimming fat occasionally. Strain liquid, and discard vegetables and wings.

Keep chicken stock, cover in refrigerator.

CRANBERRY GAZPACHO

1 cucumber

1 red pepper, cored and seeded

1 jalapeno pepper, seeded

1/2 green pepper, cored and seeded

2 tomatoes

4 scallions

1 bunch parsley

1 tablespoon lemon juice or wine vinegar

salt and pepper

1 cup water

1/2 cup fresh cranberries

Serves 4

Preparation time: 20 minutes, plus overnight chilling

Mince cucumber, peppers, tomatoes, scallions and parsley. Combine in large bowl.

Add lemon juice, salt and pepper. Set aside.

In a 1-quart saucepan, bring 1 cup of water to a boil. Cook cranberries for about 1 minute, or until a few cranberries split. Let the cranberries cool. When cool, add them to the gazpacho.

Cover and refrigerate overnight.

CHILLED RASPBERRY AND BLACKBERRY SOUP

$^{1}/_{3}$ cup red raspberries
$^{1}/_{3}$ cup blackberries
$^{2}/_{3}$ cup buttermilk
$^{2}/_{3}$ cup plain yogurt
pinch of salt

Pass raspberries through a sieve and discard the seeds. In a separate bowl, pass blackberries through a sieve and discard the seeds.

Add half the buttermilk and yogurt to the raspberries. Add half the buttermilk and yogurt to the blackberries.

Add salt to each and chill 1 hour. With a 4-ounce ladle pour the blackberry soup into a bowl, and a ladleful of the raspberry soup in the center.

Serves 2
Preparation time: 1 hour 20 minutes

FIDDLEHEAD SALAD WITH BLUEBERRY VINAIGRETTE

1/4 cup blueberry vinegar
1 tablespoon Dijon mustard
1/2-3/4 cup extra virgin olive oil
salt and pepper

1 pound fresh fiddlehead ferns
1 head red leaf lettuce
1 head radicchio
1 head bibb lettuce
4 bunches mâche, or lamb's lettuce
1 cup fresh blueberries

Serves 16
Preparation time: 20 minutes

VINAIGRETTE:

In a small bowl, whisk together the vinegar and the mustard. Add the olive oil, salt and pepper.

Drop the fiddlehead ferns into boiling water for 1 minute. Rinse under cold water. Wash lettuce thoroughly in clean water.

In a salad bowl, combine the lettuce with the ferns. Toss with the vinaigrette and garnish with the fresh blueberries.

SUN-DRIED CRANBERRY COLE SLAW

1 cup sun-dried cranberries

1 cup white zinfandel, or any blush wine

1 cup vinaigrette (see page 38)

1 cup sour cream

1 cup mayonnaise

1 small head red cabbage (about 1 pound)

2 medium-sized red onions, sliced thinly

2-3 scallions, thinly julienned

2 tablespoons ground cumin

1/3 cup sugar

salt and pepper.

Serves 10-12

Preparation time: 20 minutes, plus overnight twice

Soak cranberries in wine overnight at room temperature.

Combine vinaigrette, sour cream and mayonnaise. Set aside.

Remove outer leaves from cabbage. Core and cut into quarters. Slice thinly.

Combine onion and cabbage with sour cream mixture. Add the cranberries, scallions, ground cumin, sugar and salt and pepper. Cover and refrigerate overnight.

CHICKEN SALAD WITH BLACKBERRIES

1 frying chicken, about $3\frac{1}{2}$ pounds

salt and pepper

1 cup water

1 egg yolk

1 tablespoon Dijon mustard

$\frac{1}{2}$ cup olive oil

2 tablespoons lemon juice or white wine vinegar

$\frac{1}{2}$ cup sour cream

1 bunch radishes, quartered

1 head chicory, broken into leaves

4 scallions, cut lengthwise and into thirds

2 cups fresh blackberries, rinsed

Serves 7

Preparation time: 2 hours

Preheat oven to 400º F. Rub chicken with salt and pepper. Place chicken in a $8\frac{1}{2}$ x 11- x 2-inch roasting pan with a cup of water. Roast for 50-60 minutes. Let cool.

Pick chicken meat by removing skin and separating all meat from bones, and set aside.

In a bowl, whisk together egg yolk and mustard. Still whisking, gradually add olive oil and lemon juice or vinegar. Salt and pepper to taste. Combine the olive oil mixture with the sour cream. Toss chicken pieces with the egg-oil-sour cream mixture. Add radishes.

Lay a bed of chicory on a plate. Top with $\frac{2}{3}$ cup of chicken.

Sprinkle cut scallions over chicken. Arrange berries on plate.

SHRIMP SALAD WITH REDCURRANTS

1 lemon, halved
10 black peppercorns
2 quarts cold water
12 large shrimps
4 plum tomatoes
1/2 cup red currants
salt and pepper
4 leaves romaine lettuce

Serves 2
Preparation time: 30 minutes

Place halved lemon and peppercorns in a quart of cold water. Heat to boiling; reduce heat and simmer for 5 minutes. Place the shrimp in the water and remove from heat. Let stand for 5 minutes. Drain and let the shrimp cool. Shell and devein shrimp. Discard lemon and pepper.

In a second pot, heat to boiling second quart of water. Place tomatoes in the water and boil for about 3 minutes. Transfer tomatoes into ice water. Skin the tomatoes, cut in half, remove seeds and dice. Mix with currants in medium bowl. Add salt and pepper. Shred lettuce and arrange 2 leaves on each plate. Place tomato-currant mixture on top of the lettuce. Arrange the shrimp on top of the mixture.

ENDIVE SALAD WITH BLACKBERRIES

juice of 1 lime
2 tablespoons balsamic vinegar
2 tablespoons sherry vinegar
2 tablespoons red wine vinegar
1 egg yolk
2 tablespoons Dijon mustard
2 tablespoons honey
1 tablespoon soy sauce
1 cup peanut oil
2 tablespoons walnut oil
salt and pepper
2 heads Belgian endive
1/4 cup chopped walnuts
1/4 cup fresh blackberries

Serves 2
Preparation time: 2 hours, 15 minutes

Combine lime juice, vinegars, egg yolk, mustard, honey and soy sauce in a bowl. Whisk. Add oils, salt and pepper to taste. Cover and refrigerate for about 2 hours.

Separate the outer, larger leaves of the endive and arrange on a plate. Just before serving, cut the remaining inner leaves into 1/4-inch rings. Sprinkle walnuts and berries over the endive leaves. Drizzle about 2 tablespoons of the vinaigrette over salad. Refrigerate additional vinaigrette for later use.

BITTER GREENS SALAD WITH STRAWBERRIES AND GOOSEBERRY VINAIGRETTE

2 tablespoons gooseberry preserves
2 tablespoons rice wine vinegar
1 teaspoon Dijon mustard
1/4 cup safflower oil
1/4 cup olive oil
2 tablespoons heavy cream
salt and pepper
10-12 small cauliflower florets
1 bunch watercress
1 bunch dandelion greens
4 ounces goat cheese
1/2 cup strawberries

Serves 2
Preparation time: 1 hour

Whisk together preserves and vinegar until smooth.

Continue whisking, while slowly adding Dijon mustard and oils.

Add cream, salt and pepper, whisking gently. Refrigerate, covered, for 1 hour.

Drop cauliflower into 1 cup boiling water and boil for 1 minute. Remove florets from boiling water and run under cold water.

Arrange greens on plate with cauliflower, goat cheese and strawberries. Drizzle 2 tablespoons of dressing over salad.

SMOKED TROUT AND RASPBERRY SALAD WITH LINGONBERRY DRESSING

1 cup crème fraiche

4 tablespoons lingonberry preserve

juice of 1 lemon

salt and pepper to taste

2 bunches arugula, washed thoroughly in cold water, and torn into bite-sized pieces

1 red onion, thinly sliced

1 whole smoked trout (about 1/4 pound), picked off the bone and broken into bite-sized pieces

8 plum tomatoes, cut into quarters

1 cup fresh raspberries

Serves 4

Preparation time: 15 minutes

In a small bowl, combine crème fraiche, preserves, lemon juice, salt and pepper.

In medium bowl, toss arugula with red onion. Spread onto 4 plates. Arrange trout pieces, tomato pieces and raspberries. Add dressing.

FRUIT SALAD

2 medium-sized juice oranges

2 limes

1/4 cup raspberries

1/4 cup golden raspberries

1/4 cup strawberries

1/4 cup blueberries

1/4 cup blackberries

1 large papaya

2 kiwi fruit, peeled and sliced

1/4 cup shredded coconut

Serves 4

Preparation time: 15 minutes, plus 2 hours chilling

With a sharp paring knife, remove all skin from the oranges. Holding the oranges over a medium bowl to catch juices, cut along the membranes of the oranges so the sections, or supremes, fall into the bowl as well. Repeat the process for the limes. Add raspberries, strawberries, blueberries and blackberries. Cover and refrigerate for 2 hours.

Peel papaya, cut in half and remove the seeds. Cut each half into 1/2-inch slices and arrange on 4 plates with sliced kiwi fruit.

Spoon berry and orange mixture, with juices, over papaya. Sprinkle with coconut.

THREE
MAIN COURSES

CRAB CAKES WITH BLACKBERRIES

3-4 cups boiling water

1 cup instant grits

1/2 cup unsalted butter

1 teaspoon salt

8 ounces Maine crabmeat, large lump, rinsed and cleaned

1/2 cup fresh blackberries

5 tablespoons sour cream

5 teaspoons black lumpfish caviar

Serves 5

Preparation time: 1 1/2 hours

Add grits, 4 tablespoons butter and salt to the boiling water and stir. Lower the heat and continue stirring. When grits are smooth and have absorbed all the water, set aside and let cool.

Combine cooked grits with crabmeat. Add blackberries. Form into 10 patties and chill for 1 hour.

In a large skillet, melt remaining butter over medium to low heat. When butter begins to crackle (but not burn), add the crab cakes. Cook on each side until golden brown.

Serve each portion of 2 crab cakes with 1 tablespoon of sour cream and 1 teaspoon of caviar.

BLACKENED TUNA WITH STRAWBERRIES

1 tablespoon unsalted butter

5-6 large strawberries, stemmed and cut into thirds

1/2 teaspoon ground cumin

1/2 teaspoon ground cinnamon

1/2 teaspoon ground marjoram

1/2 teaspoon ground cayenne pepper

salt and pepper

2 8-ounce tuna steaks

4 tablespoons peanut oil

Serves 2

Preparation time: 15 minutes

In small skillet over low heat, melt the butter and add strawberries. Sauté strawberries for 3 minutes, until soft. Set aside.

Mix together cumin, cinnamon, marjoram, cayenne pepper, salt and pepper. Season tuna generously with mixture. In small cast-iron skillet, heat peanut oil over high heat. Sear fish until black, about 3-4 minutes. Turn over, and blacken other side, another 2-3 minutes. (This will produce a lot of smoke.) Garnish tuna with strawberries and serve.

STUFFED LOBSTER WITH CRANBERRIES

2 tablespoons unsalted butter

2 tablespoons all-purpose flour

1 cup half-and-half

1 gallon plus one cup water

1/2 cup cranberries

1/4 cup canned corn, drained, or fresh corn kernels

1/4 red pepper, finely chopped

1 jalapeno pepper, seeded and finely chopped

salt and pepper

1 1/2 pound Maine lobster

Serves 2

Preparation time: 1 hour

In a skillet over medium heat, melt butter and add the flour. Whisk together for 4-5 minutes to form a roux. The roux will be extremely hot. Using whisk, add half-and-half slowly. Set aside.

In a saucepan, bring 1 cup of water to a boil. Cook cranberries for about 1 minute, or until the cranberries begin to split.

Add corn, red pepper, jalapeno and cranberries to the half-and-half mixture. Add salt and pepper. Set aside.

In an 8-quart pot bring 1 gallon of water to a boil. Lower to a simmer. Put the lobster in the pot by holding the body, claws facing down, and cover for 6 minutes. Remove lobster and let stand until cool enough to handle.

Split the lobster in half lengthwise with a 10- or 12-inch chefs knife. To do this, cut from the base of the head towards the tail, with shell side up. Turn lobster around and split head. With a fork, remove meat from the tail and chop coarsely. Combine with the corn-and-pepper mixture. Stuff mixture back into the tail shell. (Can be refrigerated, covered in plastic wrap, for up to 2 days.)

Preheat oven to 425º F. Place lobster in baking dish. Bake lobster for 8-10 minutes.

GRILLED SWORDFISH WITH RED CURRANT BUTTER

½ cup water

3 tablespoons fresh red currants

¼ cup unsalted butter, softened to room temperature

salt and pepper

2 8-ounce pieces swordfish

1 tablespoon olive oil

Serves 2

Preparation time: 1 hour and 15 minutes

In a 1-quart saucepan, bring to a simmer the ½ cup water. Add the currants and poach for 1 minute. Strain currants and allow to cool. Combine currants, butter, and salt and pepper.

Form the currant butter into a 5-inch log. Place it in the butter wrapper, plastic wrap, or wax paper and refrigerate for 1 hour.

Preheat grill to high heat according to manufacturer's directions.

Season swordfish with salt and pepper. Rub with olive oil. Grill both sides of swordfish for 3-4 minutes each, until tender to the touch.

Garnish each fish with a slice of currant butter.

STEAMED SHRIMP WITH CRANBERRY COULIS

1/4 cup Cranberry Gazpacho (see page 34)

1 1/4 cups water

1/2 cup basmati rice

salt and pepper

12 medium-sized shrimp

4 tablespoons very finely chopped parsley

Serves 2

Preparation time: 1 hour, including preparation of gazpacho

Prepare Cranberry Gazpacho and set aside.

In a small saucepan, bring to a boil 3/4 cup water. Add the rice and lower the heat. Stir, while adding salt and pepper. Cover and cook for 12-15 minutes, until all the liquid is absorbed. Set aside.

In a small saucepan on medium heat, heat the Cranberry Gazpacho. Add 2 tablespoons of water and cook until the water is evaporated. Cover. Set aside.

In a deep saucepan, bring to a boil 1 cup of water. Place the shrimp in a strainer and place the strainer in the saucepan. Cover for about 3-4 minutes, or until shrimp are firm and completely white. Set aside.

Toss parsley with rice, and arrange on plate. Place Cranberry Gazpacho on the rice.

Arrange shrimp on plate.

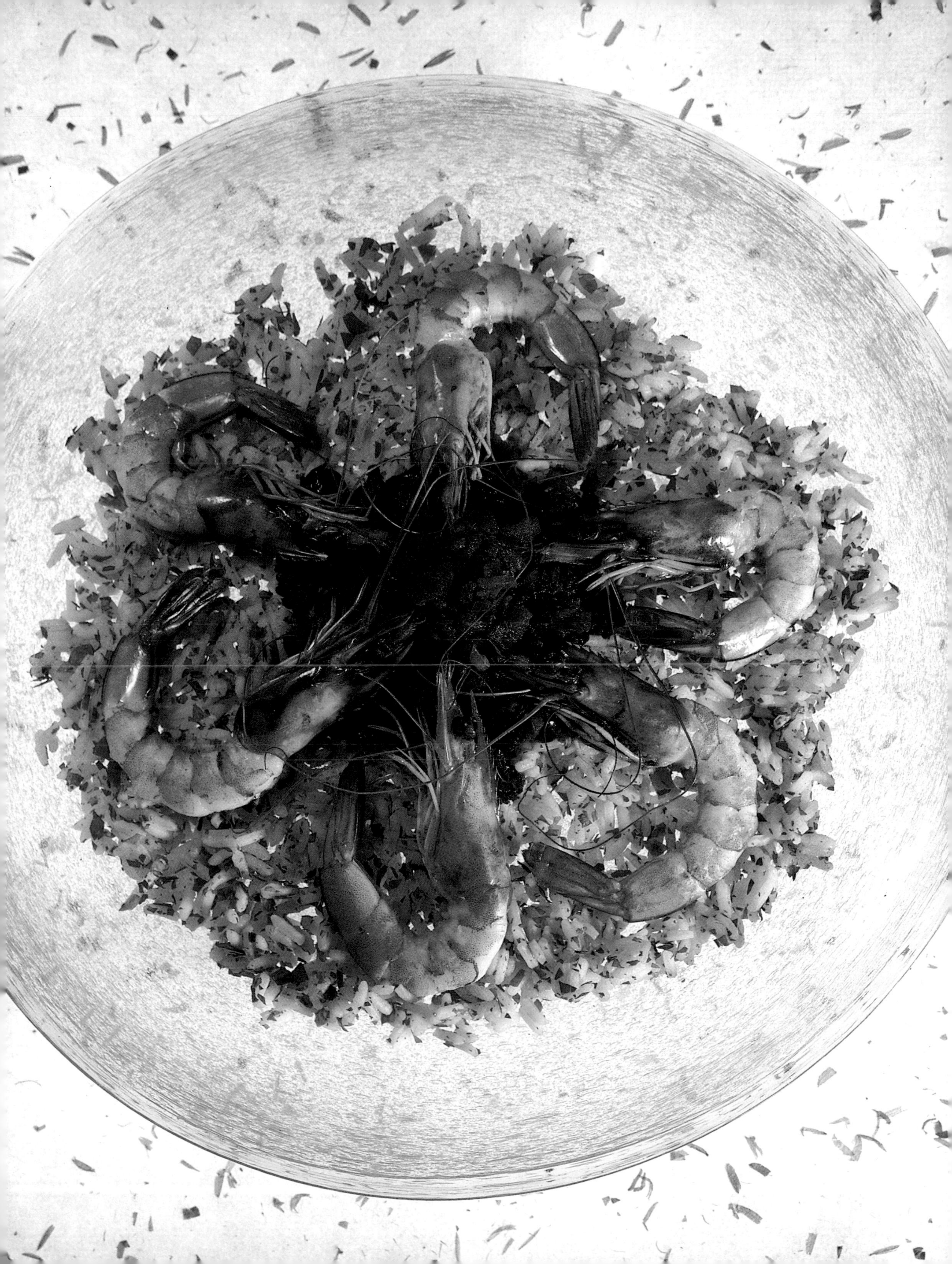

SQUID INK FETTUCINE WITH BLACKBERRY CREAM SAUCE

1 cup boiling water

4 stalks asparagus trimmed and cut in 2-inch long pieces

2 tablespoons unsalted butter

3 large shallots, peeled and sliced into rings

1/2 cup dry white wine

1/4 cup heavy cream

3 plum tomatoes, chopped

1 cup fresh blackberries

salt and pepper to taste

2 quarts water

10 ounces fresh squid ink

fettucine

Serves 2

Preparation time: 15 minutes

Blanch asparagus in boiling water for 1 minute. Remove from heat and run under cold water until cold throughout.

Set aside.

In a large skillet, melt butter and add shallots. Cook until transparent.

Add wine and bring to a boil. Allow to reduce by half. Add the heavy cream and reduce mixture by half. Lower the heat.

Add the tomatoes, asparagus, blackberries, salt and pepper.

Heat water, lightly salted, to boiling in a large pot. Cook fettucine for 3 minutes, or until done. Strain and mix the pasta with the sauce in medium bowl.

PAN-FRIED GROUPER WITH BLUEBERRY BUTTER SAUCE

- 2 8-ounce grouper fillets
- salt and pepper
- 1/4 cup flour
- 2 tablespoons peanut oil
- 1/3 cup blueberry vinegar
- 1/4 cup heavy cream
- 4 tablespoons unsalted butter, softened

Preheat oven to 425º F.

Season grouper with salt and pepper. Dust lightly with flour. Set aside. In a large skillet, heat peanut oil until it smokes. Place fish in skillet and brown on one side. Turn over and remove from skillet. Place in baking pan and bake for about 6 minutes. Remove fish from oven and return to skillet. Raise to high heat.

Add vinegar and allow liquid to reduce almost completely. Add heavy cream and bring to a boil. After cream has thickened gradually add the butter. Add salt and pepper. Place grouper on plate and spoon sauce on top.

Serves 2

Preparation time: 15 minutes

PORK RIBS WITH RASPBERRY BARBECUE SAUCE

2 racks of pork rib (about 12 bones each), skin peeled off backs

salt and pepper

1 1/2 cups water

1 cup raspberry preserves

1 cup tomato catsup

1 medium-sized yellow onion, finely chopped

1/2 teaspoon cayenne pepper

Serves 4-6

Preparation time: 3 hours

Preheat oven to 250º F. Tie the two racks together with non-coated string, so the curves of the bone face one another. Season with salt and pepper.

Stand ribs in an 11- x 14- x 2-inch roasting pan. Add 1 1/2 cups of water. Cover with aluminum foil. Roast for 2 hours, until your fingers touch through the meat when you pinch between two bones. Set aside.

In a skillet over low heat, combine preserves, catsup, chopped onion and cayenne pepper. Simmer for 5 minutes. Add salt and pepper to taste. Thin with water if sauce becomes too thick.

Cut rib racks in half, so there are about 5-6 bones per section. Cover with barbecue sauce.

ROAST LAMB WITH LINGONBERRY SAUCE

1 6-8 ounce loin of lamb

salt and pepper

2 cloves garlic, finely chopped

1 slice of bread, any type

1/4 cup sherry, brandy or red wine

1/4 cup chicken stock (see page 32)

1/4 cup lingonberries, fresh or in syrup

1 tablespoon unsalted butter

Serves 2

Preparation time: 30 minutes

Preheat oven to 450º F.

Remove all fat from meat. With a sharp knife cut off the silverskin, or shiny outer skin from the meat. Season with salt and pepper. In a very hot, ovenproof skillet, briefly sear both sides of the loin, until golden brown, over high heat on stove.

Remove from heat and rub garlic on both sides.

Place the slice of bread in the center of the skillet and top with the lamb. This elevation helps the meat to cook evenly. Roast for 5-10 minutes, until desired doneness. The handle of the skillet will be very hot. Slice lamb diagonally and arrange on plate.

Return skillet to stovetop. Over medium high heat, pour sherry into pan. Momentary flames will rise. Reduce sherry somewhat, and add chicken stock.

Bring stock to a boil and allow to reduce for 1 minute. Add the lingonberries (drain if in syrup) and butter. Swirl pan until the butter melts evenly. Add salt and pepper to taste. Pour sauce over sliced lamb.

BRAISED PORK CHOP WITH BLUEBERRY CREAM SAUCE

2 pork chops, 6-8 ounces each

salt and pepper

1/2 cup sherry

1/4 cup sun-dried blueberries

1/2 cup heavy cream

Serves 2

Preparation time: 30 minutes

Preheat oven to 400º F. Season pork chops with salt and pepper.

In a hot ovenproof skillet, sear both sides of pork chops over high heat on stove. Pour sherry around sides of skillet. Momentary flames will rise. When flames subside, add blueberries.

Cover entire skillet with aluminum foil, and put in the oven for 15 minutes, until the chops are firm to the touch.

Set the pork chops aside and return skillet to a high heat. Handle will be extremely hot. Reduce liquid by half and add cream. Bring to a boil and cook for about 2 minutes, until cream thickens. Add salt and pepper to taste.

Pour sauce over pork chops and serve.

GRILLED CHICKEN BREAST WITH GOLDEN RASPBERRY CAKES

2 cloves garlic, finely chopped

½ cup blueberry vinegar

juice of 1 lime

1½ cups virgin olive oil

1 tablespoon whole black peppercorns, crushed

2 whole boneless chicken breasts, pounded thin

1 yellow squash

1 egg

¼ cup breadcrumbs

½ cup fresh golden raspberries

salt and pepper to taste

Serves 2

Preparation time: 2 hours

In a large bowl, combine the garlic, blueberry vinegar, lime juice, 1 cup olive oil and crushed black peppercorns. Place the chicken in the marinade. Let stand 1 hour.

Grate the yellow squash into a large mixing bowl. Add the egg, breadcrumbs, golden raspberries, salt and pepper.

Form 4 patties and refrigerate 1 hour.

Set grill to high heat according to manufacturer's directions. With skin side up, place chicken on the hottest part of the grill. Turn every 1 or 2 minutes until done.

In a 12-inch skillet heat remaining oil until it smokes. Place the cakes in the pan, lower heat and cook on both sides until golden brown. Serve with Rosehip Mayonnaise.

ROSEHIP MAYONNAISE

2 egg yolks

1 tablespoon Dijon mustard

1 tablespoon lemon juice

2 cups olive oil

salt and pepper

2 tablespoons rosehip preserves

Makes 600 ml (1 pint) mayonnaise

Preparation time: 10 minutes

Combine egg yolks, mustard and lemon juice.

Whisk in the olive oil drop by drop. Add salt and pepper. When mixture is firm, fold in preserves.

CHICKEN-KUMQUAT SKEWERS WITH SPICY CRANBERRY KETCHUP

10 6-inch bamboo skewers

5 whole boneless chicken breasts

20 kumquats, left whole

1 1/4 cups fresh cranberries

1/2 cup tomato paste

1/2 cup red wine vinegar

4 jalapeno peppers, seeded and finely chopped

20 shakes Tabasco sauce

1/4 cup sugar

salt and pepper

1/ cup vegetable oil

Makes 10 skewers

Preparation time: 45 minutes

Soak bamboo skewers in water for 30 minutes to avoid burning them during cooking. Cut each chicken breast into 6 equal pieces. On each skewer alternate 3 pieces of chicken with 2 kumquats.

Bring to a boil a large pot of water. Cook cranberries for about 2 minutes, until berries begin to split. Drain.

Place cranberries, tomato paste, vinegar, jalapeno peppers, Tabasco sauce, sugar, salt and pepper in a food processor. Purée until smooth. Strain through a fine-mesh strainer.

Set grill to high heat according to manufacturers directions.

Brush each skewer lightly with vegetable oil. Salt and pepper to taste.

Place on the grill and cook until chicken is cooked throughout, turning often. Serve each skewer with 2 tablespoons catsup.

ROAST DUCK BREAST WITH PHYSALIS

2 tablespoons peanut oil

2 medium-sized duck breasts

1/4 cup white zinfandel, or other blush wine

1/4 cup chicken stock (see page 32)

1/4 cup gooseberries

1 tablespoon cold unsalted butter

salt and pepper

Serves 2

Preparation time: 25 minutes

Preheat oven to 450º F.

In a heavy ovenproof skillet, heat oil until smoking. Place duck in skillet, skin side down. Cook for about 3 minutes over high heat on stove. Turn duck over and place skillet into the oven. Roast for 8 minutes, until firm to the touch, with slight resistance. Remove skillet from oven and discard excess oil. Slice the breasts and arrange on a plate. Cover and set aside.

Return skillet to the stove. Over a high heat, add the wine. Raise to a boil and allow to reduce by about half. Add chicken stock and gooseberries. Bring to a boil and allow liquid to reduce. Add butter and salt and pepper. Pour sauce over sliced duck breasts.

HOLIDAY TURKEY WITH CRANBERRY SAUCE

1 6-pound turkey
salt and pepper
1 1/2 cups water
2 Granny Smith apples
2 sprigs fresh mint
1 stick cinnamon
3-4 whole cloves
1 1/4 cups cranberries
1/4 cup sugar or to taste
2 tablespoons unsalted butter
2 tablespoons all-purpose flour

Serves 4
Preparation time: 2 hours

Preheat oven to 375º F.

Rinse turkey thoroughly with water. Rub evenly with salt and pepper. Open bag of giblets and rinse giblets.

Place turkey in a 2-inch deep roasting pan with 1 1/2 cups of water. Add giblets to water and roast for about 1 1/2 hours or until juices run clear from the joints. Baste at 15-minute intervals. Add more water if necessary.

Peel and core apples, reserving skins. In a cheesecloth tie skins, cores, mint, cinnamon and cloves. Place apples, cranberries and bouquet garni of skins and spices in a 4-quart saucepan with enough cold water to cover ingredients. Cook over a moderately high heat for 1 hour. Stir often with a wooden spoon and add more water, if necessary. Lower the heat if necessary. When water is replenished, use cold water and bring back liquid to a boil; then lower to a simmer. This process will soften the apples and cranberries and make the sauce smoother.

Remove bouquet garni, add sugar to taste and set aside. When turkey is cooked, set it aside for 10 minutes. Strain the drippings and set aside.

To make gravy, melt the butter in a skillet and add the flour. Whisk together to form a roux. Add the drippings from the turkey pan and water, if necessary. Add salt and pepper.

Serve turkey with gravy, warm, and cranberry sauce.

FOUR
DESSERTS

LINZER BISCUITS

1 cup sugar

1 cup plus 1 tablespoon unsalted butter

1/2 teaspoon salt

1 teaspoon vanilla extract

2 cups all-purpose flour

1 cup raspberry preserves

1/2 cup confectioners' sugar

Makes 10 biscuits

Preparation time: 1 1/2 hours

Mix sugar, 1 cup butter, salt and vanilla extract until smooth. Add flour. Mix until incorporated and dough forms a ball.

Form into a log and refrigerate for 1 hour.

Preheat oven to 350º F. Cut log into 20 equal pieces. Roll each piece to about 1/4-inch thickness.

With a 4-inch ridged cookie cutter, press cookie shapes out of the rolled dough. With a 1/2-inch round cookie cutter, press a hole in the center of half the cookies.

Butter a 14- x 17-inch cookie sheet with remaining butter and place cookies at 1-inch intervals. Bake 8-10 minutes, until cookies turn golden brown. Remove from cookie sheet and cool.

Spread preserves on the halves of the cookies without center holes.

Top with the cookies with center holes.

Place confectioners' sugar in a strainer and dust over the cookies.

RASPBERRY AND BLUEBERRY TARTLETS WITH LEMON MOUSSE

LEMON MOUSSE

zest and juice of 3 lemons

3 egg yolks

3 tablespoons sugar

1/2 cup heavy cream

1/2 cup blueberries

1/2 cup raspberries

TART SHELLS

1 cup all-purpose flour

1/2 cup cold unsalted butter, cut into small pieces

1/3 teaspoon salt

2 tablespoons sugar

Serves 4

Preparation time: 3 1/2 hours

LEMON MOUSSE:

In a glass or stainless steel bowl, combine lemon zest, lemon juice, egg yolks and sugar. Place in a double boiler and stir until the mixture firms. Refrigerate for 3 hours. Mixture will continue to firm when cooled.

While mixture is in refrigerator, prepare tart shells.

TART SHELLS:

In a large bowl mix flour, butter, salt and sugar with fingertips.

Add a few tablespoons of cold water to bind the dough.

Divide the dough in quarters. and refrigerate for 2 hours.

Preheat oven to 350º F.

On a clean, flat surface, roll out dough and press into 4 4-inch tart pans. Bake for about 25 minutes, until golden brown. Let cool.

While tart shells cool, remove lemon mixture from refrigerator and proceed with lemon mousse.

In a clean bowl, whip the cream until it holds to the sides.

Mix some of the whipped cream with the lemon mixture; then fold that mixture into the rest of the whipped cream. Spread into tart shells.

Arrange blueberries and raspberries (1/4 cup berries per tart, combined or alone) on top of the mousse.

BLUEBERRY PIE

- 1 cup (2 sticks) cold unsalted butter, cut into small pieces
- 2 cups all-purpose flour
- 1/4 cup sugar
- 2 teaspoons salt
- 4-5 tablespoons ice water
- 2 cups fresh blueberries
- 1 egg
- 2 tablespoons water

Serves 6

Preparation time: 3 hours

Mix butter, flour, 1/4 cup sugar and 1 teaspoon salt with fork or fingers until mixture has the consistency of coarse cornmeal. Add ice water, 1 tablespoon at a time, until dough binds.

Divide the dough in half, wrap in plastic and refrigerate for about 2 hours.

Preheat oven to 350º F. On a clean, dry surface, sprinkle some flour and roll out one of the dough patties. Press into an 8-inch pie pan.

In a medium bowl, toss blueberries with 1/2 cup sugar and 1 teaspoon salt. Fill pie shell with blueberries.

Roll out remaining dough, and cut into 1/4-inch wide strips. Arrange on top of pie in a criss-cross pattern. Crimp edges.

Beat the egg with 2 tablespoons of water and brush over the surface of the pie.

Bake for 45-50 minutes, or until golden brown.

BLACKBERRY CUSTARD

1 teaspoon unsalted butter

1/2 cup sugar

2 egg yolks

2 cups heavy cream

1/2 cup fresh blackberries

Serves 4

Preparation time: 2 hours

Preheat oven to 350º F.

Grease 4 6-ounce cups with the butter and coat with 2 tablespoons sugar.

Combine egg yolks and 4 tablespoons sugar. Blend until mixture turns light yellow. Add cream. Skim off surface foam. Add blackberries, reserving 2 or 3 berries.

Mash reserved berries with a fork. Add remaining 2 tablespoons sugar and a few drops of water. Set aside.

Pour egg mixture into the cups. Place the cups in a 2-inch deep pan filled with enough water to reach halfway up the sides of the cups. Bake 45 minutes or until firm. Remove from pan and let cool, about 1 hour. Turn upside down onto a plate and cover with mashed berry sauce.

CHOCOLATE CAKE WITH FRAMBOISE-SOAKED RASPBERRIES

1 1/2 cup fresh raspberries

1/2 cup framboise 12 ounces semi-sweet chocolate

1 1/2 cups unsalted butter

7 tablespoons sugar

1 cup walnuts

6 eggs

2 tablespoons cornstarch, dissolved in 2 tablespoons of framboise

Serves 12

Preparation time: 3 hours, plus overnight chilling

Preheat oven to 350º F.

Soak 1 cup raspberries in framboise for 2 hours.

Over a double boiler combine chocolate, butter and 5 tablespoons sugar, and melt. Let cool.

In a blender, grind walnuts into a fine powder. Transfer to a medium bowl and add eggs, raspberries and cornstarch.

Combine walnut mixture with cooled chocolate mixture, and pour into a 12-inch springform pan.

Bake for about 40 minutes, until firm to the touch. While cake is baking, prepare raspberry sauce.

In a medium saucepan place remaining raspberries, 2 tablespoons sugar and 1 tablespoon of water. Over a low heat, cook for 10 minutes, stirring occasionally. Strain and cool. (Sauce can be refrigerated overnight.)

Refrigerate cake overnight to set. Unmold and serve with sauce.

PISTACHIO AND ORANGE SOUFFLÉ WITH HOT BLUEBERRY SAUCE

1 tablespoon unsalted butter

2 tablespoons finely chopped pistachios

1/4 cup, plus 3 tablespoons superfine sugar

1/2 cup fresh blueberries

2 eggs, separated

zest of 1 orange

2 teaspoons Grand Marnier, or other orange-flavored liqueur

Serves 2

Preparation time: 25 minutes

Preheat oven to 350º F.

Butter 2 6-ounce soufflé cups. Combine the chopped pistachios with 2 tablespoons of sugar and dust into soufflé-cups. Chill while continuing recipe.

In a medium pan over a low heat, cook the blueberries with 3 tablespoons sugar and a few drops of water. Add more water if necessary. Cook until blueberries split. Lower heat to keep blueberries warm.

Combine egg yolks, orange zest, Grand Marnier and 2 tablespoons sugar.

In deep, narrow bowl, beat egg whites until stiff peaks form. Fold some of the egg whites into the yolk mixture; then fold the egg yolk mixture into the remaining egg whites.

Using a rubber spatula, fill the 2 cups with the batter. Place the cups in a 2-inch deep pan filled with enough water to reach halfway up the sides of the cups. Bake 10 minutes, until the soufflé rises above the lip of the cup.

Spoon the blueberry sauce on top of the soufflé and serve immediately.

HAZELNUT CHEESECAKE WITH BLACKBERRIES

1 cup blackberries

1 1/2 cups sugar

1/2 cup hazelnuts

1/2 cup cream cheese

1/2 cup sour cream

1/2 cup ricotta cheese

2 eggs

pinch of salt

1/2 teaspoon vanilla extract

1/2 cup unsalted butter

1 cup all-purpose flour

Serves 6

Preparation time: 4 hours

Preheat oven to 350º F. In medium bowl, combine blackberries with 1/2 cup sugar. Cover and refrigerate 1 hour. Roast hazelnuts in a shallow pan for 5 minutes. Cool and chop coarsely.

In clean bowl, combine 1/2 cup sugar, cream cheese, sour cream, ricotta cheese, eggs, salt and vanilla. Set aside.

In another bowl, cream 1/2 cup sugar with butter. Add flour and a couple of drops of water. Mix until the dough binds. Add roasted hazelnuts.

Roll out dough and press into a 9-inch springform pan. Fill with cream cheese mixture.

Bake for 1 hour, until the cake is firm to the touch.

Chill for 2 hours.

Cover individual slices with blackberries and syrup.

RASPBERRY MOUSSE

1 cup frozen raspberries, thawed, or fresh raspberries

2 egg yolks

2 tablespoons sugar

1 cup heavy cream

zest of 1 lime

Strain the raspberries and syrup, and discard the seeds.

Add the egg yolks and sugar to the raspberries.

Place the mixture in a double boiler, and stir slowly with a rubber spatula for 10 minutes, until it begins to firm. Refrigerate for 1 hour. The mixture will become firmer as it chills. In a tall, narrow bowl, whip cream till stiff peaks form. Fold a bit of whipped cream into the raspberry mixture. Fold the raspberry mixture into the remaining whipped cream. Sprinkle lime zest on top and serve.

Serves 4

Preparation time: 1½ hours

COEURS À LA CRÈME

1/2 cup cream cheese, softened
1 teaspoon vanilla extract
2-3 tablespoons confectioners' sugar
1/2 cup heavy cream
1/2 cup fresh raspberries
2 tablespoons sugar

Serves 2

Preparation time: 15 minutes, plus overnight chilling

In a medium bowl, blend cream cheese, vanilla and confectioners' sugar until creamy. In a deep, narrow bowl, whip heavy cream until firm. Mix some of the whipped cream with the cream cheese mixture; then fold the cream cheese mixture into the remaining whipped cream. Line 2 coeurs à la crème molds* with damp cheesecloth. Fill the lined molds with the cream cheese mixture. Refrigerate the molds overnight on a plate to catch whey, or drippings.

In a medium saucepan place all but 6 to 8 of the raspberries, the sugar and 1 tablespoon of water. Over a low heat, cook for 10 minutes. Stir occasionally. Add more water if necessary. Strain and cool.

Unmold coeurs à la crème, and serve with remaining raspberries and sauce.

*These molds come in two sizes – large and small – and are available at specialty kitchenware or pottery stores.

GÉNOISE WITH LINGONBERRY BUTTER CREAM

GÉNOISE

1 teaspoon butter

1 cup all-purpose flour

6 eggs

1/4 cup sugar

pinch of salt

4 tablespoons unsalted butter, melted and cooled

LINGONBERRY BUTTER CREAM

4 egg yolks

1/2 cup confectioner's sugar

1 cup unsalted butter, softened

12 ounces lingonberry preserves

1/2 cup heavy cream

Serves 6

Preparation time: 2 1/2 hours

GÉNOISE:

Preheat oven to 350º F.

Butter 2 9-inch round baking pans with 1/2 teaspoon butter each. Sprinkle each with 2 tablespoons of flour. Discard excess flour. Refrigerate.

In a bowl, whisk whole eggs and sugar until very firm and pale.

Sift remaining flour and salt into egg-and-sugar mixture.

Fold until incorporated.

Add melted butter and continue folding.

Pour batter into cake pans. Pound the pans lightly against a table-top to eliminate large air pockets. Do not bang too hard or the batter will deflate.

Bake for 20-25 minutes, or until edges shrink away from the pan and a toothpick inserted in the center comes out clean.

Let cool. Cut each cake in half horizontally.

LINGONBERRY BUTTER CREAM:

In a bowl, blend egg yolks and confectioner's sugar until the mixture becomes a light yellow color.

Add the softened butter gradually until it is evenly incorporated. Refrigerate for 1 hour.

Add the preserves to the butter cream. Refrigerate for 1 hour.

In a deep, narrow bowl, whip the heavy cream until stiff.

Spread the butter cream evenly over each cake half. Spread the whipped cream over all but one of the halves, on top of the butter cream. The layer without whipped cream will be the top layer. Assemble cake.

BLUEBERRY SORBET

36 ounces lemon-lime flavored, carbonated soda

1 cup fresh blueberries

Serves 10

Preparation time: 3 hours

Pour soda in a 2-inch deep flat pan. Place, uncovered, in the freezer. Freeze until solid.

Remove from freezer and scrape frozen soda with the dull edge of a knife into a medium-sized bowl. Add blueberries to the slush.

Cover the mixture and return to freezer.

Freeze until firm

STRAWBERRY ICE CREAM

5 egg yolks

1 quart heavy cream

1 cup sugar

1-1½ tablespoons strawberry-flavored liqueur

½ cup fresh strawberries

Serves 10

Preparation time: 45 minutes

Combine all ingredients. Follow instructions for your ice-cream machine. Use a machine with a 1½-quart capacity. Cover and store in freezer.

CHOCOLATE-DIPPED STRAWBERRIES

4 ounces semi-sweet dark chocolate
6 strawberries

Melt the chocolate over a double boiler.

Holding the strawberries by the stem, dip each one in the melted chocolate until evenly covered. Place on wax paper and refrigerate.

Serves 2
Preparation time: 20 minutes

SUMMER PUDDING

2 cups fresh berries (raspberries, blackberries, strawberries and black currants all work well combined or alone)

1/2 cup superfine sugar

2 cups heavy cream

6 slices white bread, crusts trimmed off

Serves 2

Preparation time: 1 hour, plus overnight chilling

Toss the berries with 1/4 cup of sugar, and set aside for 1 hour. Berries will give off juice.

In a deep, narrow bowl, whip the cream to stiff peaks with the remaining 1/4 cup of sugar. Cover and refrigerate.

Cut the bread into 1-inch squares.

In 2 goblets or small dessert bowls, place several berries along with 1 tablespoon of juice per serving.

Spread 2 tablespoons of whipped cream on top of the berries.

Cover the whipped cream with a layer of bread pieces. Repeat the procedure, layering berries with juice, whipped cream and bread pieces until the bowl is filled. Cover and refrigerate overnight.

Top with the remaining whipped cream and a few berries as garnish.

FIVE
Beverages

CRANBERRY GINGER TEA

2 cups boiling water
1/2 cup fresh ginger, thinly sliced
1/2 cup fresh cranberries, rinsed
pinch nutmeg
1/2 cup cranberry juice
2 sprigs of mint

Serves 2
Preparation time: 25 minutes

In a medium-sized bowl, pour boiling water over ginger and cranberries. Cover and let stand 20 minutes. Strain, add nutmeg and cranberry juice and stir.

Serve warm or chilled over ice cubes. Garnish with mint.

BLUEBERRY YOGURT SHAKE

2 cups plain yogurt
1/2 cup orange juice, freshly squeezed
1 cup fresh blueberries, rinsed
1 banana, very ripe

Serves 4
Preparation time: 3 minutes

Combine all ingredients in a blender. Blend on medium speed until smooth and frothy.

Pour into glasses and serve.

NUTTY RASPBERRY

1/4 ounce hazelnut-flavored liqueur
1/4 ounce raspberry-flavored brandy
1/4 ounce heavy cream (optional)

Combine in a cocktail glass, over ice.
If using cream, combine in a shaker. Shake and pour into a glass, over ice.

Serves 2
Preparation time: 1 minute

FRAMBOISE AND CHAMPAGNE

1 ounce framboise, or raspberry-flavored brandy
4 ounces champagne, chilled

Pour brandy into champagne glass. Add champagne.

Serves 2
Preparation time: 1 minute

BERRY ICE CUBES

12 blueberries
12 raspberries

Fill 2 ice cube trays with 1 or 2 berries in each section. Cover with warm water and freeze overnight.

Makes 20 ice cubes, with 10-cube trays.
Preparation time: 1 minute, plus overnight freezing

STRAWBERRY MARGARITA

$1^1/_2$ pints fresh strawberries, hulled
5-6 ice cubes, crushed, roughly 1 cup
juice of 1 lime
$1^1/_2$ ounces golden tequila

In a blender, on low speed, combine all ingredients.

Serves 2
Preparation time: 1 minute

STRAWBERRY DAIQUIRI

$^1/_2$ pint fresh strawberries, hulled
5-6 ice cubes, crushed, roughly 1 cup
juice of 1 lime
$1^1/_2$ ounces light rum

In a blender, on low speed, combine all ingredients.

Serves 2
Preparation time: 1 minute

RASPBERRY PUNCH

1 quart ginger ale
1/2 cup golden rum
1/2 cup raspberry-flavored brandy
juice of 1 lime

Combine ingredients, proportionately to taste.

Serves 10
Preparation time: 5 minutes

RASPBERRY EGGNOG

1 quart heavy cream
1 quart milk
4 eggs
pinch of nutmeg
2 cups raspberry-flavored brandy
1 cup sugar
1 quart vanilla ice cream

Combine ingredients, proportionately to taste.

Serves 15
Preparation time: 5 minutes

*For both recipes, the amounts and proportions ought to be determined by personal taste. Begin with all ingredients listed, and taste as you gradually add each ingredient.

RECIPE LIST

Recipe List by berry

Gooseberry

Lingonberry

Mulberry

Raspberry

Other titles available in this series